Canon EOS Rebel T7 User Guide

A Step-by-Step Canon Rebel T7 User Guide for Beginners, Seniors, and Aspiring Photographers — Master Camera Settings, Manual Mode, HD Video, and Stunning Photography with Ease

Georgette Howard

1

Table of Contents

11

Introduction

Welcome to Your New Creative Journey

So, you've just unboxed your Canon EOS Rebel T7 sleek, powerful, and full of possibilities. But let's be honest: if you're anything like most new camera owners, the buttons feel intimidating, the manual feels overwhelming, and you're not quite sure where to begin.

That's exactly why this book exists.

Why This Guide?

The Canon EOS Rebel T7 is one of the most beginner-friendly DSLR cameras ever made but if you've held it in your hands and felt a bit overwhelmed by the buttons, settings, and unfamiliar terms, you're not alone.

This guide was created for real people not just photographers or tech enthusiasts. Whether you're holding

13

a DSLR for the first time or returning after years away, this book is your simple, stress-free path to confidence. No overcomplicated language. No unnecessary theory. Just what you *actually* need to start shooting incredible photos today.

You don't need to know everything to take great pictures. You just need someone to walk you through the right things at the right time.

That's exactly what this guide does.

Who This Book Is For

This book is designed for:

- **Beginners** who want to understand their Rebel T7 without flipping through a boring manual
- **Seniors** or less tech-savvy users who want things explained in plain English

- **Hobbyists and casual shooters** who want to improve their skills and take better photos

- **Parents, travelers, bloggers, or small business owners** who want high-quality images without hiring a pro

- **Anyone switching from smartphones** and curious about unlocking the power of a DSLR

If you've ever said, "I just want to take great pictures without the tech headaches," then this guide is for you.

What You'll Be Able to Do by the End

By the final chapter, you'll know how to:

- ✓ Confidently set up, hold, and operate your Canon EOS Rebel T7

- ✓ Understand and use all the basic and advanced shooting modes

- ✓ Capture clear, sharp photos in any environment both indoors or outdoors
- ✓ Master beginner-friendly manual settings for portraits, landscapes, and action shots
- ✓ Transfer, edit, and share your photos like a pro without needing expensive software
- ✓ Avoid common mistakes and troubleshoot camera problems with ease
- ✓ Start building a creative portfolio that reflects your unique perspective

You'll go from wondering **"How do I even use this camera?"** to proudly saying **"I took that!"**

Chapter 1

Meet Your Canon EOS Rebel T7

Get Comfortable with Your Camera — One Step at a Time

Before we dive into settings, modes, and creative techniques, let's slow down and really get to know your camera. Because the more familiar you are with your Rebel T7, its buttons, its layout, and its parts the faster everything else will start to make sense.

This chapter is your first hands-on experience with the camera. Let's unpack it together.

Unboxing and What's in the Box

When you open your Canon EOS Rebel T7 package, you should find the following items:

- Canon EOS Rebel T7 Camera Body

- EF-S 18–55mm Kit Lens (if you purchased the standard bundle)

- Battery Pack LP-E10

- Battery Charger LC-E10

- Neck Strap (EW-400D)

- Camera Interface Cover

- Printed Quick Start Guide or Instruction Manual

- Warranty Card

Tip: Keep the box and internal packaging, they're great for safe transport or resale later.

The Anatomy of the T7: Buttons, Ports, and Dials Explained

At first glance, your camera may seem like a maze of buttons. Don't worry most of them have just one or two functions, and you'll quickly develop muscle memory.

Here's a breakdown of what you're looking at:

Top of the Camera

- **Mode Dial** – Where you switch between Auto, Manual (M), Portrait, Landscape, and more

- **Shutter Button** – Press halfway to focus, fully to take a picture

- **Power Switch** – Turns the camera ON and OFF

- **Main Dial** – Used to adjust settings like aperture and shutter speed in Manual mode
- **Flash Button** – Pops up the built-in flash
- **Hot Shoe Mount** – For external flash or accessories

Back of the Camera

- **Viewfinder** – For composing your shot with your eye (great in bright light)
- **LCD Screen** – Where you review photos, change settings, and navigate menus
- **Menu Button** – Opens the settings menu
- **Playback Button** – View your photos
- **Q (Quick Control) Button** – Quick access to settings
- **D-Pad and Set Button** – Navigate and confirm selections

Side Ports

- **USB/AV Port** – For connecting to a computer or TV

- **HDMI Port** – For external display connections

- **Mic input?** – Unfortunately, the Rebel T7 doesn't support external mics.

Charging, Inserting the Battery & Memory Card

Before anything else, give your battery a full charge.

To charge the battery:

1. Insert the **LP-E10 battery** into the **LC-E10 charger**

2. Plug the charger into a wall outlet

3. The light will turn orange while charging, and green when fully charged (about 2 hours)

To insert the battery and memory card:

1. Flip the camera upside down and open the battery

compartment door

2. Insert the battery (contacts facing inward) until it clicks

3. Insert a Class 10 SD card (at least 16GB recommended) next to the battery

4. Close the latch securely

Pro Tip: *Format the memory card before your first use. (Menu → Setup → Format Card)*

Holding Your Camera the Right Way

Proper handling isn't just about comfort — it directly affects how sharp your photos are. Shaky hands = blurry images.

Here's how to hold the Rebel T7 like a pro:

- Right hand grips the camera body with your index finger on the shutter button

- Left hand cradles the lens from underneath (not the side)

- Keep elbows tucked into your body for support

- Stand with one foot slightly ahead of the other for balance

Steady shot = sharper image. Always use two hands and good posture.

Quick Recap

- Know what came in your box and check for missing items

- Familiarize yourself with the layout and purpose of each button

- Fully charge and install your battery and memory card

- Practice your grip and handling for sharper, cleaner shots

Chapter 2

Getting Set Up the Right Way

Start Smart So You Can Shoot With Confidence

Now that you're familiar with the buttons, battery, and how to hold your Rebel T7, it's time to turn it on and set things up the right way from the very beginning. These few small steps can make a big difference later from smoother shooting to easier photo sharing.

Let's walk through it all no tech stress, just simple wins.

GETTING SET UP THE RIGHT WAY
• Setting Date, Time, and Language
• Navigating the Menu with Ease
• Connecting to Wi-Fi and Bluetooth
(and Why It Matters)
• Installing the Canon Camera Connect App

Setting the Date, Time, and Language

The first time you turn your Canon Rebel T7 on, the camera will prompt you to set the Date/Time and Language.

Here's how to do it:

1. Turn the **Mode Dial** to the green Auto mode (for now).

2. Flip the **Power switch** to **ON.**

3. The **Date/Time** setting screen will appear.

4. Use the **arrow keys** (D-pad) to select your values:

- Set the **year**, **month**, **day**, **hour**, and **minute**.

- Press the **SET** button to confirm each.

5. Next, you'll be prompted to set the **Language** — select your preferred option (e.g., English) and press **SET** again.

Why this matters: Your camera will tag each photo with the correct time/date super helpful when organizing, uploading, or printing images.

Navigating the Menu with Ease

The Rebel T7's menu may look intimidating at first, but it's actually clean and logical once you understand the layout.

To open the menu:

- Press the MENU button (top-left of the screen when looking from the back).

To navigate:

- Use the **left/right** arrows to move between different tabs at the top.
- Use the **up/down** arrows to move through the options within each tab.
- Press **SET** to select or change a setting.

Menu Tabs Overview:

1. **Shooting Settings (red tabs)** – Where you'll adjust photo & video options.
2. **Playback Settings (blue tab)** – Review/delete images.

3. **Setup Settings (yellow tab)** – Format SD cards, date/time, LCD brightness, etc.

4. **Display Settings (purple tab)** – Customize how your screen looks.

5. **My Menu (green tab)** – Create shortcuts to your favorite settings.

Tip: If something confuses you, just press MENU again to exit safely.

Connecting to Wi-Fi and Bluetooth (and Why It Matters)

Even though the Rebel T7 is a "starter" DSLR, it still has built-in Wi-Fi and that opens up awesome, time-saving features:

- Transfer photos wirelessly to your phone
- Remote shoot using your smartphone

- Instant social media uploads without needing a computer

How to set up Wi-Fi:

- Press **MENU** and go to the Setup tab (yellow).
- Scroll to **Wi-Fi/NFC** → Press **SET** → Choose **Enable**.
- Now go to **Wi-Fi Function** → Choose **Connect to Smartphone.**
- You'll see the camera's Wi-Fi name and password.
- On your smartphone, go to Wi-Fi settings and connect to that network.
- Then open the **Canon Camera Connect app** (see below).

Installing the Canon Camera Connect App

The **Canon Camera Connect** app is your camera's perfect companion. It allows you to:

30

- Preview shots live on your phone

- Transfer images quickly to your gallery

- Control the shutter remotely (perfect for group shots or selfies)

How to install:

- On iPhone: Open the App Store → Search "Canon Camera Connect" → Download.

- On Android: Open Google Play → Search "Canon Camera Connect" → Download.

Once installed:

- Open the app and follow the on-screen instructions to pair with your camera.

- Use the app to remotely capture images, view photos, or download them directly to your phone.

Security Tip: You can disable Wi-Fi/NFC when not in use to save battery and avoid accidental connections.

Quick Setup Recap:

- Set **date**, **time**, and **language** for proper image tracking.

- Learn the **menu layout** — it's your command center.

- **Connect to Wi-Fi** for faster photo sharing.

- Download and link the **Canon Camera Connect app** to simplify your workflow.

Chapter 3

Understanding Photography Basics

Unlock the Power of Your Camera by Learning the Language of Light

Every great photo starts with one thing: light. But the magic of photography is more than just pointing and shooting, it's about understanding how your camera sees light and how you can control it.

In this chapter, we'll break down the core principles of exposure, focus, file formats, and shooting modes in a way that finally makes sense, no jargon, no math degree required.

By the end, you'll stop guessing and start creating with confidence.

What Is Exposure? Aperture, Shutter Speed, ISO Made Simple

Exposure is how bright or dark your photo appears. Your camera controls this using three settings — often called the Exposure Triangle:

Aperture (The Lens Opening)

- Think of it like the pupil of your eye — it opens or closes to let in more or less light.
- Measured in f-stops (like f/2.8, f/4, f/5.6, f/8...)
- Lower f-stop = brighter image + blurrier background (great for portraits)
- Higher f-stop = darker image + more in focus (great for landscapes)

Example: f/2.8 = bright with blurry background

f/11 = darker but sharp from front to back

Shutter Speed (How Long the Sensor Is Exposed)

- Measured in fractions of a second (e.g., 1/1000, 1/250, 1/30…)
- Faster speed = frozen action (sports, wildlife)
- Slower speed = motion blur or low-light help (waterfalls, night)

Tip: Use a tripod for slow shutter speeds under 1/60 to avoid blur.

ISO (Camera Sensor Sensitivity)

- Higher ISO = brighter images but also more noise (grain)
- Lower ISO = cleaner image, but needs more light

Best Practice: Use the lowest ISO possible for clean images (like ISO 100 or 200)

Together, aperture, shutter speed, and ISO form your creative toolkit. Learn to balance them, and you'll control how your images look and feel.

Autofocus vs Manual Focus

Your Rebel T7 is built to make focusing easy, but it helps

to know when to trust auto — and when to take over manually.

Autofocus (AF):

- Best for everyday shooting
- Press the shutter halfway to lock focus
- Works fast in good lighting and with contrast-rich subjects

Manual Focus (MF):

- Turn the switch on your lens to MF
- Twist the focus ring to sharpen manually
- Best for macro, low light, or when autofocus hunts

Pro Tip: Use Live View and zoom in on the LCD for precise manual focusing.

Shooting in JPG vs RAW

This choice affects **how your photos are saved**, and it's crucial depending on how you plan to use them.

JPG (JPEG)

- Smaller file size
- Pre-processed and ready to share immediately
- Less flexible for editing

Best for: everyday shooting, quick social media posts, casual use

RAW

- Uncompressed image with all original data
- Requires editing with software (like Lightroom or Canon's DPP)
- Offers much more control over exposure, white balance, and color correction

Best for: professional editing, print work, high-quality results

Change this in Menu → Image Quality → Choose RAW, JPG, or RAW+JPG

Camera Modes Explained: Auto, P, Tv, Av, M, and More

Your Rebel T7 includes several shooting modes — some automatic, some semi-automatic, and some fully manual. Here's what they all mean:

Auto (Green Mode)

- Let the camera handle everything, great for total beginners or quick snapshots.

P (Program Mode)

- The camera sets shutter and aperture, but you control ISO, focus, flash, etc.

Tv (Shutter Priority)

- You set the shutter speed, camera adjusts aperture.

Best for: moving subjects, action shots

Av (Aperture Priority)

- You set the aperture, camera adjusts shutter speed.

Best for: controlling background blur or landscape depth

M (Manual Mode)

- You control everything: aperture, shutter, ISO.

Best for: full creative control, learning exposure, and mastering the craft

Creative Auto (CA), Portrait, Landscape, Sports, etc.

- Scene-based modes, handy for fast results tailored to your subject.

Quick Recap:

- **Exposure = Light** (Balance aperture, shutter speed, ISO)
- **Autofocus is easy**, but manual gives you total precision
- **JPG = ready to share, RAW = ready to edit**
- Learn to use **Tv, Av, and M modes** — they unlock real creativity

Chapter 4

Shooting Your First Photos

Let's Get Snapping — No Guesswork, Just Great Shots

You've set up your camera, you understand the basics, now it's time to start taking actual photos. This chapter is where theory becomes action.

We'll begin with Auto Mode, your camera's smart assistant, and use that as a launchpad to help you build solid skills in focus, framing, and sharpness.

Let's make your first shots not just possible but awesome.

Using Auto Mode Like a Pro

Your Canon EOS Rebel T7's Auto Mode (the green rectangle on the Mode Dial) is designed to take all the guesswork out of shooting. It's perfect for getting started but you can still shoot like a pro by understanding what's happening behind the scenes.

In Auto Mode:

- The camera decides aperture, shutter speed, ISO, and white balance
- Autofocus is fully active
- The flash may pop up automatically in low light
- You only need to worry about framing, timing, and focus

Pro Tip: Just because it's automatic doesn't mean you can't be intentional. Great photos still come from great decisions, like where you stand, what you shoot, and how you compose the frame.

Shooting Your First Photos

Using Auto Mode Like a Pro

- Auto camera sets a pperature, shutter-speed, ISO, and white balance
- Flassh may pop upp in low light
- Use comframe.] timing, and focus

Framing, Focusing, and Capturing Sharp Images

> Framing Your Shot

Use the rule of thirds: focus
Keep breathing room

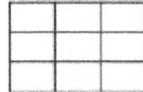

Leave breathing room in frame

> Focusing Tips

Press the shutter halfway to lock focus

> How to Keep Your Images Sharp

Hold your camera correctly ⟍ Use faster sraseeds
Use good during good li

How to Avoid Blurry or Washed-Out Photos	Washed-Out (OVERXQPOSED) Fixes
> Blurry Photo Fixes	Avoid direct midday sunlight
Use both hands & stable stance	Avoid direct Q] beduce Exposure Compensation if needed
Don't move camera until after the click	Use flash sparningly indoors
	The "Quick Shot" Checklist for Beginners

Framing, Focusing, and Capturing Sharp Images

A well-composed photo grabs attention, tells a story, and feels intentional. Let's break that down.

Framing Your Shot

44

- Use the **Rule of Thirds**: imagine your screen divided into a 3x3 grid.

- Place your subject near one of the intersections not always in the center.

- Leave breathing room, don't cut off heads or feet unless it's stylistic.

Focusing Tips

- Press the shutter halfway to lock focus

- You'll hear a beep and see a red light flash in the viewfinder

- Then press it down fully to capture the shot

Use the center focus point for best accuracy, aim it at the subject's eyes (for people) or the most detailed area.

How to Keep Your Images Sharp

- **Hold your camera correctly:** two hands, elbows tucked in

45

- **Use faster shutter speeds** in low light (Auto will try, but movement can still sneak in)

- **Light helps focus:** brighter environments = clearer images

How to Avoid Blurry or Washed-Out Photos

Even in Auto Mode, bad lighting or camera shake can still lead to poor shots. Here's how to dodge the most common mistakes:

Blurry Photo Fixes:

- Use both hands and a stable stance
- Don't move the camera until after the shutter clicks
- Avoid zooming in too far, use your feet to get closer instead

Washed-Out (Overexposed) Fixes:

- Avoid direct midday sunlight — it's too harsh

46

- Tap **Q (Quick Settings)** and reduce **Exposure Compensation** slightly if needed

- Use **flash sparingly** indoors — it often flattens the image

Golden Hour Tip: Shoot outdoors just after sunrise or before sunset — the light is soft, flattering, and magical.

The "Quick Shot" Checklist for Beginners

Before pressing the shutter, run through this mental checklist, it'll save you from 90% of common mistakes:

- ➢ Is your **subject well-lit** (natural light or bright area)?

- ➢ Is your **focus locked** on the right spot (eyes, detail)?

- ➢ Is your camera **steady** (both hands, no shake)?

- ➢ Is the background **clean** (no clutter, distractions)?

➢ Did you give your subject **space** to breathe in the frame?

If you can check these five things — even in Auto — you're going to see a major jump in quality.

You Did It!

You've taken your first real shots, not just random clicks, but intentional photographs. Whether you're capturing your kids, your pets, nature, or a favorite place, you now have the foundation to shoot with purpose and polish.

Chapter 5

Unlocking Scene Modes & Creative Filters

Instant Creativity, No Manual Mode Required

Now that you're confident with Auto Mode and basic framing, it's time to explore the built-in scene modes and creative filters on your Rebel T7, your camera's hidden treasure chest of artistic shortcuts.

With just a twist of the Mode Dial, you can shoot beautifully exposed, genre-specific photos without needing to touch ISO, aperture, or shutter speed. Let's discover what each mode is for, and how to use them to create images with personality, style, and emotion.

Portraits, Landscapes, Close-Ups — Which Mode to

Use and When

Your Canon Rebel T7 includes Scene Intelligent Auto modes, each tailored for a specific type of photography. These are marked with icons on the Mode Dial, like a face, mountain, flower, etc.

Portrait Mode (Face icon)

- Best for: People, selfies, headshots, kids
- Blurs the background to make your subject pop
- Uses wide aperture for a soft, creamy look

Tip: Use a plain background and natural light for stunning portraits.

Landscape Mode (Mountain icon)

- Best for: Nature, cityscapes, wide shots with lots of detail
- Increases depth of field — keeps everything sharp

- Boosts greens and blues for rich outdoor tones

Tip: Try shooting during golden hour (just after sunrise or before sunset) for magical lighting.

Close-Up Mode (Flower icon)

- Best for: Plants, textures, jewelry, food, small objects
- Sharpens the center and softens the background
- Boosts contrast and color saturation

Tip: Get as close as your lens allows — but let autofocus do the work.

Sports Mode (Running person icon)

- Best for: Moving subjects — kids, pets, sports, vehicles
- Uses fast shutter speed to freeze motion

- Prioritizes continuous autofocus for sharp action shots

Tip: Hold down the shutter for burst mode to capture multiple frames.

Night Portrait / Handheld Night Scene

- Best for: Low-light or nighttime scenes with people
- Slows shutter and boosts ISO to expose dark scenes
- Flash may fire — hold steady!

Tip: Use a tripod or steady surface for sharper results.

Using Creative Auto (CA) Mode for Artistic Shots

CA Mode (Creative Auto) is a smart hybrid: it gives you the simplicity of Auto, but lets you tweak the mood, brightness, and background blur without diving into settings.

To use CA Mode:

- Turn the Mode Dial to CA

- On the screen, use the SET button and arrows to:

- Adjust background blur (more or less)

- Make the image brighter or darker

- Choose ambience: vivid, soft, warm, cool, etc.

CA is like a training ground for Manual Mode — you make creative choices, and the camera handles the rest.

Fun with Filters: Toy Camera, Miniature, Vivid & More

The Rebel T7 includes Creative Filters that can turn everyday scenes into stylized, eye-catching images — all right inside your camera.

To apply filters:

1. Switch to Live View (press the LV button)

2. Turn the Mode Dial to Creative Filters

3. Scroll through options like:

Toy Camera Effect

- Adds heavy vignette and retro color tones

- Great for vintage, moody street scenes

Miniature Effect

- Blurs top and bottom for a "tilt-shift" toy town look

- Fun for shooting from above

Vivid

- Boosts color saturation and contrast

- Ideal for festivals, flowers, or food

Soft Focus

- Blurs slightly for a dreamy, romantic look

- Perfect for portraits or soft mood shots

Grainy Black & White

- Adds classic film-style grain and contrast
- Excellent for dramatic storytelling shots

Tip: Filters only apply to JPG files, not RAW so be sure to shoot JPG when using them.

Quick Recap:

- Use **Scene Modes** to get pro-level results in specific situations — no manual settings required
- Try **CA Mode** for gentle creative control while staying safe in Auto
- Explore **Creative Filters** to add mood and style directly in-camera — no editing needed

Chapter 6

Mastering Manual Mode (Without the Headache)

Full Control, One Simple Step at a Time

Manual Mode often feels like the final frontier for beginners — mysterious, risky, maybe even a bit scary. But here's the truth: it's easier than you think, and once you learn how to balance just three settings, you'll never look at photography the same way again.

This chapter walks you through the Exposure Triangle, real-life examples, and ready-to-use beginner settings so you can shoot in Manual without guesswork.

Step-by-Step: Shooting in Manual Mode

To enter Manual Mode:

- Turn the Mode Dial to M

You now have full control of:

- **Aperture** (how much light passes through the lens)
- **Shutter Speed** (how long the sensor is exposed)
- **ISO** (how sensitive the sensor is to light)

Use the Main Dial (next to the shutter button) to adjust settings, the back screen and viewfinder will preview how bright or dark your photo will be.

Don't panic, if your image looks too dark or bright, just adjust one of the three settings!

Balancing Exposure Triangle — A Simple Formula

Let's revisit the Exposure Triangle with a practical mindset. You don't need to memorize terms, just understand their effect.

Aperture (f/number)

- Low number (e.g. f/3.5) = more light, blurry background
- High number (e.g. f/11) = less light, sharper background

Shutter Speed (1/xxx sec)

- Fast (e.g. 1/500) = freezes action, less light
- Slow (e.g. 1/30) = motion blur, more light

ISO

- Low (100–400) = clean image, more light needed
- High (800–3200) = brighter image, but adds grain

Your goal is to balance all three so your meter (bottom of screen) is centered.

Real Examples: Daylight, Indoor, and Night Shoots

Let's see Manual Mode in action with real-world examples:

Bright Daylight (Outdoors)

- Aperture: f/8
- Shutter Speed: 1/400
- ISO: 100

Use for landscapes or wide shots in sunshine.

Indoor with Window Light

- Aperture: f/4
- Shutter Speed: 1/100
- ISO: 400–800

Use near a window or bright lamp, hand-held.

Night Street or City Lights

- Aperture: f/2.8
- Shutter Speed: 1/30 (use tripod if possible)
- ISO: 1600+

Use for nighttime ambiance and creative light trails.

Start with these, then adjust one setting at a time while watching your exposure bar.

Cheat Sheet: My Favorite Beginner Manual Settings

Use this chart as a mental shortcut whenever you're unsure where to start:

Subject	Aperture	Shutter Speed	ISO
Portraits (Blur Bkg)	f/2.8–f/4	1/125+	100–400
Landscapes	f/8–f/11	1/200+	100
Action/Sports	f/4–f/5.6	1/1000+	400–800

Indoors (still)	f/2.8–f/5.6	1/60+	800–1600
Low Light (tripod)	f/2.8	1/15–1/30	1600+

Reminder: Use the light meter bar (in viewfinder or LCD) as your guide. Aim for the center.

Final Thoughts on Manual Mode

Manual Mode is not about being technical, it's about being intentional. You tell the camera what you want, and it does exactly that. With a little practice, you'll find Manual Mode to be freeing, fast, and full of creative potential.

You don't need to shoot in Manual every time, but once you understand it, you'll always have the power to take control when it counts.

Chapter 7

Exploring Canon Rebel T7 Lenses

The Right Lens Can Change Everything

Your camera is powerful but it's the lens that shapes the way the world appears through your viewfinder. Whether you're capturing crisp portraits, distant wildlife, or artistic close-ups, choosing the right lens can transform your images from average to extraordinary.

This chapter walks you through lens basics, practical lens choices, and how to safely handle your gear — so you can shoot with clarity and creative freedom.

Understanding Kit vs Prime vs Zoom Lenses

Before diving into which lens to buy, it's important to understand the three basic types of lenses you'll come across:

Kit Lens (e.g., EF-S 18–55mm f/3.5–5.6)

- Comes bundled with most entry-level Canon cameras

- Offers decent flexibility for everyday use: landscapes, portraits, snapshots

- Not great in low light, but solid for beginners

Best for: all-purpose shooting, learning camera basics

Prime Lens (e.g., Canon EF 50mm f/1.8)

- Fixed focal length (no zoom), but much sharper and brighter

- Wide aperture = great for portraits and low light

- Lightweight and affordable

Best for: stunning portraits, blurred backgrounds, indoors

Popular nickname: the "nifty fifty" (50mm f/1.8)

Zoom Lens (e.g., Canon EF-S 55–250mm f/4–5.6)

- Variable focal length — lets you zoom in or out without changing lenses

- Great for shooting sports, wildlife, or distant subjects

- Often heavier and slower in low light

Best for: action, nature, travel, sports

Best Starter Lenses for Portraits, Nature, and Action

If you're ready to go beyond the kit lens, here are some game-changing options to consider:

For Portraits (People, Pets, Products):

- Canon EF 50mm f/1.8 STM – incredible bokeh, sharp focus, super affordable

- Canon EF-S 24mm f/2.8 STM – pancake lens, great for travel portraits & indoor shots

For Nature and Landscapes:

- Canon EF-S 10–18mm f/4.5–5.6 IS STM – wide-angle lens for dramatic landscapes
- Canon EF-S 18–135mm f/3.5–5.6 IS – versatile walkaround lens with good range

For Action and Wildlife:

- Canon EF-S 55–250mm f/4–5.6 IS STM – budget telephoto lens for fast-moving subjects
- Canon EF 70–300mm f/4–5.6 IS II USM – higher quality zoom for serious reach

Don't rush to buy everything. One well-chosen lens can unlock your style.

How to Change Lenses Safely

Changing your lens properly keeps your sensor clean and your gear protected.

Step-by-Step:

1. Turn off the camera.

2. Hold the camera face down (to prevent dust from falling in).

3. Press the lens release button and twist the lens counter-clockwise.

4. Align the white (EF-S) or red (EF) dot on the new lens with the dot on the camera.

5. Twist clockwise until it clicks.

Extra Tips:

- Always cap both ends of the lens not in use.

- Use a clean, dry microfiber cloth to wipe lens surfaces gently.

- Store lenses in a dry, padded bag to prevent scratches or mold.

Lens Terminology Decoded

Here's what those mysterious lens specs actually mean:

Term	What It Means
EF / EF-S	EF = full-frame; EF-S = for crop sensors like the Rebel T7
f/1.8, f/3.5	Aperture — lower = more light, blurrier background
IS	Image Stabilization — reduces blur in handheld shots
STM / USM	Focus motor type — STM = smooth/silent (great for video), USM = fast/accurate

mm (millimeter)	Focal length — lower = wide view, higher = zoomed-in field

Example: Canon EF 50mm f/1.8 STM = prime lens, 50mm field of view, bright aperture, smooth focus.

Quick Recap:

- Kit lenses are great for learning, but prime and zoom lenses unlock specialized creativity.
- Choose your next lens based on what you shoot most, not what looks cool.
- Learn to change lenses confidently and keep your gear clean.
- Don't get stuck in the gear race — the right lens is the one that matches your story.

Chapter 8

Take Control with Custom Settings

Personalize Your Camera, Simplify Your Shooting

By now, you've learned how to capture sharp photos, switch lenses, and even shoot in Manual Mode. But there's still a whole world of customization that makes your Canon EOS Rebel T7 work exactly the way you want it to.

This chapter walks you through easy-to-use settings that enhance your shooting experience, from fine-tuning light sensitivity to enabling burst shots and creating your own camera shortcuts.

Let's turn this camera into your creative sidekick.

Adjusting ISO, White Balance, and Metering Modes

ISO Settings (Sensitivity to Light)

Even outside of Manual Mode, you can manually control ISO for better results.

To change ISO:

1. Press the **ISO** button on top of the camera.
2. Use the **Main Dial** to scroll between options (ISO 100–6400).
3. Press **SET** to confirm.

Pro Tip: Use low ISO (100–200) for outdoor light. Use higher ISO (800–1600) for indoors or night.

White Balance (WB) Settings

White balance helps your camera understand the color of the light in your environment so whites stay white, not blue or yellow.

To change WB:

- Press the **Q button**, then scroll to **White Balance**.
- Choose from options like:

  - AWB – Auto (default)

  - Daylight, Shade, Cloudy, Tungsten, Fluorescent, Custom

Use "Cloudy" or "Shade" to warm up photos; "Tungsten" to remove yellow in indoor lighting.

Metering Modes (How Your Camera Reads Light)

Metering affects how your Rebel T7 determines the proper exposure.

To adjust:

1. Press **MENU**

2. Go to **Shooting Settings tab**

3. Select **Metering Mode**

Types:

- **Evaluative (default)** – good for most scenes

- **Center-weighted** – prioritizes center of frame

- **Spot metering** – precise control (great for portraits or backlit shots)

Using Drive Modes for Burst Shots and Self-Timers

Drive modes control what happens when you press the shutter.

To access:

- Press the **Q button** → scroll to Drive Mode

Options:

- **Single Shot** – one photo per click

- **Continuous Shooting** – hold shutter for burst shots (great for sports)

- **Self-Timer (2s / 10s)** – set delay before photo is taken

Use burst mode for action, and self-timer for selfies or tripod shots.

Custom Buttons & Shortcuts — Personalize Your Camera

Want quicker access to your favorite settings? Let's tweak how your camera behaves.

To customize buttons:

- Go to **MENU** → **Custom Functions (C.Fn)**

- Assign a button to change ISO, AF, or other settings on the fly

You can also enable **Quick Settings (Q button)** on your display:

- Live view → press **Q** → scroll and adjust ISO, WB, Focus mode, etc.

The more you use shortcuts, the less you dig through menus — saving you time and battery.

Formatting Your SD Card & Backing Up

Always format a new memory card in-camera before use, and make a habit of regular backups.

To format:

Press **MENU**

Navigate to the **Setup tab (yellow wrench icon)**

Select **Format Card** → OK

Warning: Formatting erases everything on the card, make sure your files are backed up first!

Backing up your photos:

- Transfer to computer via USB or card reader
- Or wirelessly send to your phone using **Canon Camera Connect**
- Use Google Drive, Dropbox, or external hard drives for safe storage

Regular backups protect your memories and keep your SD card healthy.

Quick Recap:

- Control ISO and White Balance for better light accuracy
- Use Drive Modes for faster or timed shots
- Personalize your setup with custom buttons and quick menus

- Regularly format and back up to keep everything running smoothly

Chapter 9

How to Shoot Better Videos with the T7

From Snapshots to Cinematics — Unlock the Video Power of Your Camera

You might think of your Canon Rebel T7 mainly as a photography tool, but it also has a surprisingly solid video mode, capable of recording Full HD (1080p) footage perfect for vlogs, interviews, family moments, or YouTube content.

This chapter guides you through the process of recording smooth, clear video without the technical confusion, so you can press record with confidence and start telling your story in motion.

Shooting HD Video with Confidence

To switch to video mode:

1. Press the **Live View (LV)** button (near the viewfinder) to activate the LCD screen.
2. Move the **Mode Dial** to the **video icon** (a camera symbol).
3. Press the **Start/Stop** button (next to the viewfinder) to begin recording.

By default, the Rebel T7 records:

- **Full HD 1920x1080p**
- At **30 frames per second** (fps), smooth enough for everyday shooting

Tip: Keep the camera still while recording, moving the camera too much while shooting can cause shaky footage.

Video Settings That Matter Most

Let's simplify the settings that actually make a difference:

Resolution & Frame Rate

You can shoot at:

- **1920x1080** (Full HD) at 30fps — Best overall choice

- **1280x720** (HD) at 60fps — Smoother motion, but slightly lower resolution

- **640x480** — Only for low-quality, small file size needs

Change these in **Menu** → **Movie Rec. Size**

Exposure Settings (Auto vs Manual)

By default, the T7 sets exposure automatically. But for more control:

79

Use **Manual Exposure for Video:**

- Go to **Menu** → **Movie Exposure** → **Manual**
- You can now adjust **shutter speed, aperture, and ISO** like you would for photos

For cinematic motion, try using a shutter speed of 1/60 at 30fps

Focus

- **Autofocus works in video**, but it's **slower and can hunt** (especially in low light).
- **Manual Focus** often gives smoother, more predictable results especially for talking head videos or slow pans.

*Use **Live View zoom** (magnify button) to check focus before you hit record.*

Tips for Smooth Handheld Footage

Shaky video is one of the fastest ways to ruin otherwise great content. Here's how to get smoother shots — even without a gimbal or tripod:

✅ Steady Shooting Techniques:

- Tuck in your elbows and keep your core steady
- Hold the camera close to your body, not outstretched
- Use a neck strap for tension support
- Move slowly — pan or tilt gently

✅ Use a Tripod or Monopod:

- For interviews, timelapses, or fixed shots, a tripod is ideal
- Lightweight monopods are great for travel or events

✅ Use the Right Lens:

- STM (Stepping Motor) lenses are quieter and smoother for video

- Avoid zooming in/out while filming unless you want that effect

Audio Tips & External Mic Options

The Canon T7 does NOT have a mic input, which means you're limited to the built-in microphone — and it's not ideal.

Workarounds:

- Use your **smartphone** as a dedicated audio recorder — apps like "Voice Record Pro" or "Easy Voice Recorder" work great

- Use an **external audio recorder** like the Zoom H1n

- **Clap once** at the beginning of your video to help sync sound in editing

Always record in a quiet space with minimal background noise.

Pro-Level Bonus Tips:

- Pre-focus before hitting record to avoid awkward hunting
- Record 10 seconds longer than needed — gives you room to edit
- Shoot B-roll — extra footage of your scene to cut between moments
- Edit your videos using free tools like iMovie, DaVinci Resolve, or Canva Video Editor

Quick Recap:

- Switch to Live View + Movie Mode to record
- Use 1080p at 30fps for sharp everyday video
- Keep movement smooth and focus steady
- Record clean audio separately for better quality

- Learn to edit and piece it all together, that's where the magic happens!

Chapter 10

Transferring, Editing & Sharing Your Photos

What Happens After the Shot is Just as Important

You've taken the shot. Now what?

For many beginners, the real magic and frustration starts after pressing the shutter. From getting images off your camera, to polishing them for social media, printing, or family albums, this chapter walks you through the entire process, step by step.

No fancy software required. No technical overwhelm. Just clean, confident image handling made easy.

How to Transfer Photos to Your Phone, Tablet, or

Computer

Your Rebel T7 gives you two primary ways to transfer images:

Option 1: Transfer via USB to Computer

- Turn off the camera and insert the USB cable (included with your camera).
- Plug the other end into your computer.
- Turn on the camera.
- Your computer should detect the camera like a USB drive.
- Open the folder and drag your images to your hard drive.

Pro Tip: Organize folders by date or event for easy tracking.

Option 2: Transfer Wirelessly Using Canon Camera

Connect App

If your T7 has Wi-Fi:

1. Enable **Wi-Fi/NFC** in the Setup Menu

2. Connect your camera to your smartphone's Wi-Fi

3. Launch the **Canon Camera Connect** app

4. Select photos to transfer directly to your phone or tablet

Best for: Instagram posts, quick shares, or mobile editing

Free & Easy Editing Tools (No Photoshop Required)

You don't need Photoshop or a subscription to make your photos shine. These free and **beginner-friendly** tools are more than enough:

Mobile Editing Apps

- Snapseed (iOS/Android): Pro-level features in a simple app
- Lightroom Mobile (Free version): Great for exposure and color edits
- VSCO: Stylish filters and vintage aesthetics

Desktop Software

- **Photoscape X** (Windows/Mac): Quick, easy edits + batch processing
- **Canva**: Add text or designs for social sharing or prints
- **GIMP**: Free Photoshop alternative (more advanced)

Start with basics: Brighten → Crop → Sharpen → Save

Instagram-Worthy Shots: Crop, Brighten, Sharpen

Want your shots to stand out online? Here's the 3-step formula used by most pros:

1. Crop Creatively

- Use the Rule of Thirds
- Eliminate distractions
- Zoom in on emotion or detail

2. Brighten and Balance

- Adjust exposure, contrast, and white balance
- Add a touch of warmth for more inviting tones
- Avoid over-editing, keep it natural

3. Sharpen Gently

- Enhances details without making it look fake

- Best applied in small amounts, especially to eyes and edges

Upload with a compelling caption, your image tells a story!

Printing Your Photos or Making Photo Books

Ready to hold your memories in your hands? It's easier than ever.

Home Printing Tips:

- Use photo paper (glossy or matte)
- Set printer to "Best Quality" or "Photo Mode"
- Print from high-resolution JPGs (not low-res previews)

Photo Book & Print Services:

- Shutterfly – photo books, calendars, gifts
- Mixbook – beautiful, customizable photo albums

- Snapfish – quick prints, posters, and mugs

- Printique or Mpix – high-quality pro lab prints

These platforms let you upload directly from your phone or computer and ship to your door.

Quick Recap:

- Transfer photos via USB or Wi-Fi with the Canon Camera Connect app

- Use free apps like Snapseed or Lightroom Mobile to enhance your images

- Crop, brighten, and sharpen — that's the magic trio for share-worthy edits

- Easily print or create photo books with online tools

Chapter 11

Troubleshooting & Maintenance

Keep Your Camera Happy, Healthy, and Problem-Free

Even the best cameras hit a few bumps along the way. Maybe your Rebel T7 isn't focusing properly. Maybe the screen froze mid-shoot. Or maybe you're just wondering how to keep it clean and working for the long haul.

Good news: most problems have quick, simple fixes and a little routine care can dramatically extend your camera's lifespan.

This chapter gives you the tools, confidence, and habits to troubleshoot like a pro and keep your gear in top shape.

Common Beginner Mistakes (And How to Fix Them

Fast)

1. Blurry Photos

Likely Causes:

- Slow shutter speed
- Motion blur
- Missed focus

✅ **Fixes:**

- Use faster shutter speed (1/125+ for handheld shots)
- Hold the camera properly with both hands
- Make sure your focus point is on the subject's face or eyes

2. Overexposed or Washed-Out Images

Likely Causes:

- Bright sunlight

- Incorrect metering or exposure compensation

✅ **Fixes:**

- Shoot in shade or use exposure compensation to darken image
- Use a lower ISO (100–200 in daylight)
- Try switching to Av or P mode for better control

3. Dark Photos Indoors

Likely Causes:

- Low light without flash
- High f-stop (narrow aperture)

✅ **Fixes:**

- Use a wider aperture (f/3.5–f/4)
- Raise ISO to 800–1600
- Try turning on the built-in flash or shoot near a light source

What to Do if Your Camera Freezes or Won't Focus

If the Camera Freezes or Becomes Unresponsive:

- Turn it **OFF** and remove the battery
- Wait 10 seconds, reinsert battery
- Turn it back **ON**

Still stuck? Format the memory card or reset all camera settings in the menu.

If the Camera Won't Focus:

- Make sure the lens is set to AF, not MF (switch on lens barrel)
- Tap the shutter halfway — listen for focus beep
- Ensure there's enough light and contrast in the scene

The camera struggles in very low light or if the subject lacks contrast (e.g. plain wall). Try aiming at an edge or turning on more light.

How to Clean Your Camera and Sensor Safely

Keeping your gear clean improves performance and image quality. But don't overdo it — gentle, regular care is enough.

Cleaning the Camera Body:

- Use a microfiber cloth to wipe smudges
- Clean the viewfinder and screen gently
- Use a soft brush or blower to remove dust from crevices

Cleaning the Lens:

- Use a lens blower or brush first (never wipe dust directly)
- Then gently wipe with lens cleaning tissue or solution

Never use glass cleaner or tissue paper, it can scratch the lens coating.

Cleaning the Sensor:

If you notice dark spots that don't wipe off the lens, it may be sensor dust

Steps:

- **Go to Menu → Sensor Cleaning → Clean Now**
- For deeper cleaning, choose **Manual Clean**, remove the lens, and use a **blower only**
- **Never touch the sensor**, if dust remains, take it to a professional

Sensor cleaning is delicate — when in doubt, leave it to the pros.

Extending the Life of Your Camera

Just like any tool, your Rebel T7 will last longer if you

treat it right. Follow these habits:

Best Practices:

- Turn off the camera before removing the lens, battery, or SD card
- Store it in a dry, padded bag (moisture = mold and rust)
- Keep extra batteries on hand and avoid draining them to 0%
- Use a screen protector and lens caps at all times

Environmental Tips:

- Don't leave your camera in a hot car or in direct sunlight
- Avoid rain or dusty environments or use a protective cover

A well-maintained camera can easily last 8–10+ years, and often becomes your go-to creative companion for life.

Common Camera Mistakes
(And How to Fix Them Fast

Common Beginner Mistakes
(And How to Fix Them Fast)

Blurry Photos
Causes: Faster speed,
umtion blur, missed focus
Fix: Use a faster shutter speed
or check focus point

Overexpos Images
Causes: Bright
exposure too high or ISO
Fix: Lower the exposure
compensation or ISO

Dark Photos Indoors
Causes: Far sine
the ISO or widen aperture
Fix: Raise the ISO or widen
the aperture (lower f-stop)

Keep capped
when not in use –
dust is the enemy!

What to Do if Your Camera
Freezes or Won't Focus

Restart: Turn the
camera off, then on

Check focus mode
(AF / MF), clean the lens

Camera and Lens
Use a blower for dust
and wipe gently with cloth

Sensor Use auto clean
in menu. Don't touch it
with your fingers!

Extending the
Life of Your
Camera

Avoid moisture
and extreme heat

Turn off before
swapping lens or
battery

Use a padded bag
for storage and
transport

Recharge batteries
regularly

Quick Recap:

- Fix blurry, dark, or overexposed photos with quick setting changes

- If the camera freezes, restart and reset

- Clean gently and regularly — sensor care is serious

99

- Protect your gear from moisture, heat, dust, and accidents

Chapter 12

Growing Beyond Auto — Learn Like a Pro

From Taking Pictures to Thinking Like a Photographer

At this point, you're no longer a beginner you've shot in Manual Mode, explored lenses, mastered the basics, and handled challenges with confidence.

Now it's time to shift from just using your camera... to thinking like a photographer.

In this final chapter, we'll help you develop real-world skills, build creative discipline, and understand light, composition, and growth. Whether your goal is to turn pro, build a hobby, or capture meaningful moments, this is your path forward.

How to Practice Like a Photographer

The biggest difference between casual shooters and skilled photographers? Intentional practice.

Here's how to build that into your routine:

1. Pick One Setting to Practice at a Time

- One day: only shoot in **Av Mode**
- Another: practice only **manual focus**
- Another: experiment with **white balance presets**

When you isolate a skill, you learn it faster and deeper.

2. Create a Daily or Weekly Photo Ritual

- Shoot **at the same time** each day (sunrise, lunch break, golden hour)
- Choose a recurring subject: your pet, street scenes, a local tree, your coffee mug

Consistency beats randomness. And you'll start seeing things others don't.

3. Review and Reflect on Your Shots

- Ask: What worked? What didn't? Why?
- Look for patterns: Are your shots underexposed? Are the backgrounds messy?

Every misstep is a mini masterclass in disguise.

Shooting Challenges to Build Confidence

Push your creativity with simple challenges designed to stretch your eye and confidence:

Weekly Shooting Prompts:

- "One Lens Only" Day – Choose one lens and shoot everything with it

- "Monochrome" Day – Look for only one color all day and photograph it creatively
- "10 Shots Only" – Take only 10 photos the entire day. Be intentional.
- "No People" Challenge – Capture stories without including faces
- "Opposite Hand" Challenge – Yes, shoot using your non-dominant hand

Constraints fuel creativity. Give your eye a reason to explore.

How to Read Light and Compose Better Images

No matter what camera or lens you own, your best images will always come down to light and composition.

Understanding Light:

- **Golden Hour (sunrise/sunset)** = warm, soft light = dreamy shots

- **Midday Sun** = harsh shadows = best for contrast or dramatic silhouettes

- **Overcast Skies** = even lighting = great for portraits and macro

Tip: Always ask yourself — "Where is the light coming from?"

Improving Composition:

- Use the **Rule of Thirds** — subject off-center = more natural balance

- Use **leading lines** — roads, fences, arms — to draw eyes into the photo

- Use **negative space** to let your subject breathe

- **Frame your subject** — shoot through windows, arches, branches

Great photos are built by where you stand and how you see, not just what you shoot.

When to Upgrade Gear (and What Comes Next)

Your Rebel T7 is a fantastic foundation. But when your skills outgrow it, here's what to consider next:

When to Upgrade:

- You **feel limited** by ISO in low light
- You want **faster autofocus** or **more frames per second**
- You crave **better video quality** or **4K resolution**
- You shoot professionally and need **dual card slots**, **weather sealing**, or **deeper customization**

What to Upgrade First:

- **Lenses** — unlock more creative control before switching bodies

106

- **Tripod or gimbal** — essential for stable video and long exposures
- **External mic or lighting** — for polished video/audio
- **Camera body** — when you've outgrown your current limitations

Beginner to Pro Progression:

- Canon Rebel T7 → Canon 90D → Canon R50 → Canon R6 Mark II (Mirrorless)
- Or jump to **full-frame mirrorless** with Canon EOS RP or R5

But remember: the best upgrade is the one you'll use often.

Final Takeaway:

Becoming a better photographer isn't about gear or perfection, it's about paying attention, practicing with purpose, and enjoying the process.

You now have the skills, the mindset, and the tools to shoot with confidence, whether you're capturing everyday life, chasing golden hour, or building your own portfolio.

Bonus Chapter

Rebel T7 Photography Recipes & Inspiration

Step-by-Step Settings, Real-Life Scenarios, and Where to Grow Next

You've mastered your camera. You've practiced your skills. Now let's bring it all together with real-life photography recipes, clear, repeatable steps you can use in the field today.

In this bonus chapter, we'll give you go-to settings for popular scenes, walk through real-world shooting examples, and help you connect with a global community of photographers who can inspire and support your growth.

Let's get practical, creative, and connected.

Step-by-Step Settings for Portraits, Landscapes, Night Scenes & More

Use these as starting points, then tweak them to match your lighting, lens, and subject.

Portraits (Indoors or Outdoors)

- **Mode:** Av (Aperture Priority)
- **Aperture:** f/2.8 – f/4 (for background blur)
- **ISO:** 100 (outdoors) / 400–800 (indoors)
- **Focus:** Face or eyes (use center focus point)
- **White Balance:** Daylight or Cloudy

Shoot in open shade or during golden hour for the best light.

Landscapes

- **Mode:** Av
- **Aperture:** f/8 – f/11 (for depth and sharpness)

- **ISO:** 100

- **Focus:** One-third into the scene (for foreground & background focus)

- **White Balance:** Daylight

Use a tripod if shooting at sunrise/sunset or in low light.

Night Scenes (City Lights, Stars, Fireworks)

- **Mode:** M

- **Aperture:** f/2.8 – f/4

- **Shutter Speed:** 1–10 seconds (use tripod!)

- **ISO:** 800–1600

- **Focus:** Manual focus, set on a bright spot or infinity

- **White Balance:** Tungsten (for cooler tone)

Turn off image stabilization on a tripod. Use 2-sec timer to reduce shake.

Action/Sports

- **Mode:** Tv (Shutter Priority)

- **Shutter Speed:** 1/1000 or faster

- **Aperture:** Auto

- **ISO:** 400–800

- **Drive Mode:** Continuous shooting

- **Focus:** AI Servo (tracking moving subjects)

Pan with your subject for dynamic motion shots.

Food or Product (Flat Lay / Close-Up)

- **Mode:** Av

- **Aperture:** f/2.8 – f/5.6

- **ISO:** 100–400

- **Focus:** Manual or spot AF

- **Lighting:** Soft natural light near a window

Use a white foam board or reflector for even fill lighting.

Real-Life Shooting Scenarios & Sample Shots

Let's walk through how a real shoot might unfold using your Rebel T7.

Scenario: Family Picnic in the Park

- Light changes from shade to sun → switch between Cloudy and Daylight WB
- Group shot → use 10-sec timer, f/5.6, ISO 200
- Candids of kids playing → switch to Tv Mode, 1/1000 shutter
- Landscape of sunset over the trees → Av Mode, f/11, ISO 100

One afternoon. Four very different scenes. One camera. Just smart choices.

Scenario: Rainy Street Reflections at Night

- Use Manual Mode: f/2.8, 1/10 sec, ISO 1600

113

- Handheld shots? Lean against a wall or pole for stability

- Reflections pop in puddles, shoot low, use streetlights for backlight

Bonus: Try Black & White filter for dramatic mood

Scenario: Indoor Mood Portraits

- Shoot near a window with sheer curtains

- Av Mode, f/2.8, ISO 800

- Use a reflector (even a white towel) to bounce soft light

- Ask your subject to look into the light, not directly at the camera

The magic isn't the gear. It's the light, the angle, the moment.

Join the Canon Photography Community

Photography is more fun and more meaningful when you're not alone.

Where to Share, Learn, and Grow:

- Canon USA Learn & Explore → learn.usa.canon.com
- Canon Photography Facebook Groups
- Reddit → r/CanonPhotography or r/photography
- Instagram → Follow hashtags like #RebelT7, #CanonPortrait, #CanonBeginner

Start Your Photo Journal or Online Portfolio:

- Use **Instagram, Flickr**, or **YouPic** to track your growth
- Challenge yourself to a **"365 Project"** — one photo a day for a year

- Build a **simple photography** blog with WordPress or Squarespace

🗨 Ask questions. Post your photos. Give feedback. Grow.

Final Thoughts:

Photography isn't about having the best camera. It's about noticing life more deeply, capturing moments more intentionally, and telling stories with every click.

With your Canon EOS Rebel T7, you've learned the how. With your curiosity and practice, you'll now grow the why.

So go shoot. Go explore. Go see the world with new eyes — one frame at a time.

Acknowledgement

To every aspiring photographer and filmmaker who dares to pick up a camera and tell a story, this book is for you. Special thanks to my family and friends for their encouragement, and to the creative community whose passion inspires me daily. Your support made this guide possible.

www.ingramcontent.com/pod-product-compliance
Lightning Source LLC
Chambersburg PA
CBHW031901200326
41597CB00012B/504